Biology of the peregrine and gyrfalcon in Greenland

William A. Burnham and
William G. Mattox

MEDDELELSER OM GRØNLAND, BIOSCIENCE 14 · 84

Table of contents

Dedicated to the memory and friendship of Finn Salomonsen

Accepted 1983
ISBN 87-17-05221-1
ISSN 0106-1054
Printed in Denmark by AiO Print as, Odense

Biology of the peregrine and gyrfalcon in Greenland

William A. Burnham and William G. Mattox

Burnham, William A. and Mattox, William G. 1984. Biology of the peregrine and gyrfalcon in Greenland. – Meddr Grønland, Biosci. 14, 25 pp. Copenhagen 1984-05-30.

A ten year study began in 1972 in West Greenland to investigate the breeding biology of the peregrine falcon. Data on nesting gyrfalcons were also collected. Thirty-four peregrine nesting sites were examined in the 6050 km^2 inland study area near Søndre Strømfjord. Limited research also centered in Disko Bugt and Frederikshåb. Peregrines were found nesting predominantly on high, south-facing cliffs which overlooked large areas. The mean minimum distance between peregrine eyries was 7.7 km for the inland area (1972 and 1973) and 55 km for the coast (1974). Approximately 60 percent of the inland nesting sites were occupied each year. A ten-year average production of 1.90 young per occupied site and 2.78 young per successful site was determined. Lapland longspurs, snow buntings, wheatears, and redpolls comprised 90 percent of the peregrine's diet. Raven nests and prey availability may affect gyrfalcon nesting. Gyrfalcons and peregrines did not breed successfully on the same cliffs as they do in Alaska where prey species number and density is greater. Competition for nest sites probably occurs, but prey availability may be the most significant factor affecting falcon density. Addled peregrine eggs, eggshell fragments, and peregrine prey species were collected. Whole eggs averaged 14.3 ppm wet weight (305 ppm lipid weight) DDE, while eggshell measurements showed a 16 percent thinning compared with pre-1940 eggs from Greenland. Prey species carried low levels of DDE. The peregrine population appears to be at a near critical contamination level, and a small increase in DDE level could contribute to a population decline. No indication of a decline has been observed during the study, and the population appears stable. The project banded 185 peregrines, from which 8 recoveries occurred. The recoveries suggest peregrines migrate south to winter in South America.

William A. Burnham, Cornell University, The Peregrine Fund, 1424 N. E. Frontage Road, Fort Collins, Colorado 80524, U.S.A.
William G. Mattox, Ohio Department of Natural Resources, Fountain Square, Columbus, Ohio 43224, U.S.A.

Introduction

For over 900 years, from the period of Viking supremacy in Europe, the great "White Falcon" of Greenland has been sought after and admired. The white Greenland falcon or gyrfalcon *(Falco rusticolus)* became known as the King's Ransom or Emperor's bird in falconry, with inestimable value placed on it. Until 1972, however, little attention focused upon the other falcon which nests in Greenland, the peregrine falcon *(Falco peregrinus)* (Figure 1). The ten year study, 1972–1981, was to determine the reproductive status, pollutant levels, density, nesting requirements, prey species, and interspecific competition of peregrine falcons and gyrfalcons in Greenland.

The peregrine falcon is smaller than the gyrfalcon and shows considerably more contrast in plumage between immature and adult. Gyrfalcons, however, vary in color from almost pure white to quite dark. Some authorities even divide the gyrfalcon into two races, both of which occur in Greenland (Salomonsen 1950–51).

Peregrines and gyrfalcons differ in ways other than color and size. The peregrine's methods of hunting and flight contrast from the gyrfalcon's (Dementiev 1960; Fischer 1968). The Greenland peregrine migrates long distances south each year, while the gyrfalcon is only slightly migratory.

During the last twenty years marked declines in peregrine falcon populations have been witnessed in many parts of the world (Hickey 1969). In recent years certain races of the peregrine have been listed as endangered. Gyrfalcons have not shown similar declines, but local breeding populations fluctuate in numbers between years with their prey species – a common phenomenon in the Arctic.

The peregrines of Greenland have been included in

Fig. 1. An adult female peregrine falcon flying above the researchers.

the arctic North American race *F. p. tundrius* (White 1968). This subspecies is considered endangered because of nesting population declines. Banding and migration data for the race as a whole probably do not support the endangered status (pers. comm. F. P. Ward), although the arctic peregrine population has declined in eastern Alaska and western Canada (Fyfe et al. 1976).

The peregrine's decline has been attributed to changing climate (Porter and White 1973), human disturbance, and the introduction of chlorinated hydrocarbon chemicals into the environment (Ratcliffe 1970). The first two factors probably have not affected the arctic peregrine. The world-wide use of chlorinated hydrocarbons, mainly DDT, poses the single greatest threat to the peregrine. Even peregrines in the relatively unpolluted Arctic are exposed to chemical pollutants by migratory prey and their own annual flights south to the wintering range.

Small birds eaten by peregrines and gyrfalcons in the Arctic migrate south to areas where agricultural chemicals are commonly used. These small birds accumulate low levels of DDT and other chemicals which are stored in fat tissues. When falcons eat such birds, they gradually accumulate chlorinated hydrocarbons in their bodies. Most tundra peregrines are thereby exposed to pesticides year around. Peregrines can even accumulate levels high enough to result in death (Prestt 1965). In most cases, however, lethal levels are never reached; instead, sub-lethal levels cause eggshell thinning and egg breakage occurs (Cade et al. 1971), a major factor in the decline of some peregrine populations (Hickey and Roelle 1969). Thickness of eggshells of peregrines in western United States decreased by 20% since DDT was introduced (Burnham et al. 1978). Organochlorine contamination also causes embryonic mortality and behavior abnormalities (Ratcliffe 1970; Newton and Bogan 1974). No eggshell thinning or related reproductive failures have been observed in gyrfalcons.

Discovery of eggshell thinning and decline of nesting peregrines in the early 1950's in Great Britain (Ratcliffe 1980) and the 1960's in North America spurred research to determine the worldwide status of the species, but no research began in Greenland until the summer of 1972 (Mattox et al. 1972; Mattox 1975; Burnham et al. 1974).

Description of study area

Inland

We established the study area in the widest portion of ice-free land of West Greenland near the Arctic Circle (Figure 2). The area is 110 km long (49°55′–52°20′ W) from the edge of the inland icecap toward the outer coast, and is 55 km wide (66°45′–67°15′N). This area of 6050 km^2 is mountainous, with elevations up to 810 m above sea level. The many mountains and valleys are divided by Søndre Strømfjord in an area dotted with nearly a thousand lakes (Figure 3). Four glacial outflow rivers traverse the area. The treeless tundra vegetation consists predominantly of sedges and grasses with low willow brush *(Salix)* and dwarf birch *(Betula)* that seldom reach 1 m in height.

A continental climate determines two features of the inland study area: large annual temperature extremes and low precipitation. Temperatures are slightly warmer inland than along the coast in summer and colder during winter when temperatures drop to –50°C. The mean annual precipitation of 15 cm falls mostly as rain during June, July, and August.

Peregrines arrive from mid-to-late May and depart in the latter half of September. Temperatures during the summer months usually remain above freezing, although temperature records collected near sea level do not represent the entire area. The mean summer temperature varies considerably with an increase in elevation. When temperatures of 8–10°C occurred at sea level, snow accumulated at elevations above 450 m.

Mountain ranges in the northern portion of the study area trend northeast-southwest while those in the southern sector trend northwest to southeast. Glacial erosion and weathering have produced steep cliff faces reaching 180 m in height.

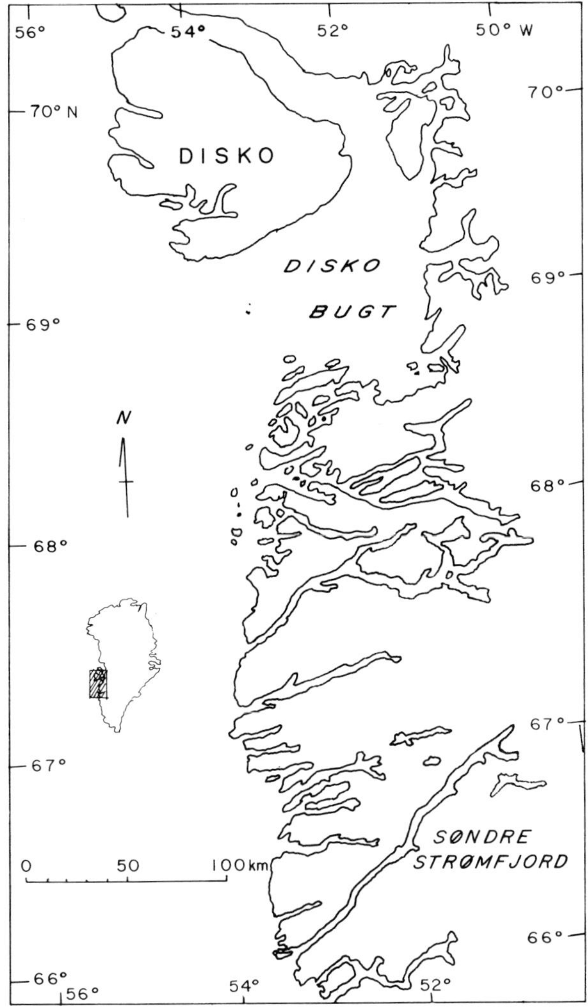

Fig. 2. Locations in Greenland where research was conducted.

A total of 29 species of birds has been seen within the study area (Table 6). Only five species of mammals have been seen. These include musk ox *(Ovibos moschatus),* harbor seal *(Phoca vitulina),* arctic fox *(Alopex lagopus),* caribou *(Rangifer tarandus),* and the smallest mammal, the arctic hare *(Lepus arcticus).* No other small mammals occur naturally in the area.

Coastal

The outer coast survey concentrated in the Disko Bugt area, 69°15′–70°15′ N, about 300 km north of the inland survey area (Figure 2). Here the summer weather is much more severe than inland; many days of rain and cold occur each spring and summer. The bay is cluttered with icebergs; frequently fjords and waterways entering the bay are not passable because of ice jams. Areas of the northern part of the bay may also be impassable during most of the summer due to ice. Disko, in the upper portion of the bay, extends slightly west beyond the mouth into Davis Strait. The island, with an area of 5578 km^2, has a central icecap and fjords penetrating the perimeter (Figure 4). Many active glaciers surround the bay. Jakobshavn Isfjord, which empties into the bay and generally is considered the world's most productive glacier, moves at a rate of 25–30 m a day.

The maritime climate of the coastal areas produces much more precipitation, including snow. We found tundra plant life in the Disko area to be much more diversified than near Søndre Strømfjord. The high mountains, alluvial plains, tidal areas, gullies and valleys of Disko help create a diverse environment for plant life. Bird life is also more diverse on the coast, with many sea birds nesting on Disko and along the surrounding coast. Large colonies of fulmars *(Fulmarus glacialis),* kittiwakes *(Rissa tridactyla),* and other birds can be located. Shorebirds are also more numerous on the coast with many sandpipers and other shorebirds frequenting alluvial areas. Most of the Disko Bugt shoreline, however, does not have concentrations of shorebirds. Passerines are few. The Lapland longspur *(Calcarius lapponicus),* the predominant inland passerine, occurs only rarely in many coastal areas because of the harsh, wet weather conditions (Salomonsen 1950–51). Snow buntings *(Plectrophenax nivalis)* are the most numerous of all passerines.

We saw fewer numbers of land mammals in the Disko area than inland. Few caribou exist, and even fox and hare numbers are reduced. Only polar bears *(Thalarctos maritimus)* are seen more frequently on the coast than inland, but they occur only rarely in West Greenland.

Method of survey

Inland

We learned of two peregrine nesting sites from residents at Søndrestrømfjord Air Base. They saw gyrfalcons frequently, but knew of only two nest sites. At the beginning of the survey we could not find small aircraft to locate nesting habitat, so we found nests by searching on foot during one- to ten-day backpacking trips. From 1973 through 1981, the survey used small aircraft to locate new cliffs, but occupancy at all sites had to be checked on foot. Helicopters were not readily available and observation from fixed-wing aircraft has not proved completely reliable.

We used boat transportation when feasible and available. Interested Danish citizens supplied rides on the fjord, and we used a small rubber raft for crossing rivers, and a folding Klepper kayak to survey the shores of large inland lakes. Unlike in Alaska, we could not float rivers to locate nesting sites because of width,

Fig. 3. The inland area as shown in this figure is more rolling than coastal areas (Fig. 4).

shallowness, quicksand, and, in some cases, unsurpassable rapids. We contracted for limited helicopter support to cross rivers and establish supply caches.

We checked all cliffs located to determine occupancy. When we found falcons we climbed the cliffs to determine nesting progress. If we found incubation in progress we withdrew from the area and did not return until the young reached over a week in age.

Coastal

Travel along the coastline required a boat to move researchers from one area of ice-free land to the next. Boats included kayaks, Zodiac boat, and Greenlandic fishing vessels. Transportation proved to be a problem because of availability, weather, and ice. We found survey work of this type slow, costly, and difficult.

Native hunters were contacted in many towns and villages in an attempt to locate eyries. Because of the limited amount of ice-free land, and many fjords with easy accessibility from the sea, the natives are quite aware of nesting locations. We found no fixed-wing aircraft available on the coast. We followed inland procedures for locating eyries and, if falcons had been reported in the area, we spent additional time attempting to locate the birds.

Peregrine biology

Nesting requirements

Height of cliff

Peregrines may nest on cut banks or on low, sandy mounds in the Canadian Arctic (Hickey and Anderson 1969). We did not find this to be true in West Greenland. Of the nesting attempts witnessed, all but one were on cliffs of from 27 m to over 117 m in height, not including steep talus slopes below most cliffs. We found only one nest on one of many low, broken cliffs. We

Fig. 4. A section of shore line on Disko where falcons have been occasionally observed by local inhabitants. The mid-elevation cliff probably has the greatest potential for nesting.

found most inland sites on predominant cliffs overlooking substantial areas. However, in 1971 a native hunter reported that a peregrine produced young on a low, inland, north-facing escarpment. We checked the area in 1972 and during the following seasons, but found no falcons. Neither did we see orange lichen *(Caloplaca)* and excreta ("whitewash") which would have indicated previous nesting. Three small outcrops existed in the area, but a recent dirt and rock slide covered most of one location. It is possible that the eyrie had been on the destroyed cliff. In another part of the survey area our team in 1979 found a pair of peregrines nesting on a low, fragmented cliff only 6 m above the ground.

When at eyries the adults frequently use their superior elevation to initiate attacks on prey (Harris and Clement 1975). The nesting cliff not only provides a launch location for attacks, but frequently has updrafts which peregrines ride to gain elevation before moving away from the cliff.

Height and inaccessibility of the inland cliff offer protection from the many arctic foxes (Figure 5). Other than man and a very occasional polar bear, the fox is the only mammalian predator which could possibly feed on young peregrines. Man can not be discounted as a predator since young falcons are considered palatable to Greenlanders. We know of eyries where young have been taken for food; however, most young appear to be killed for target practice, despite local protection laws.

All peregrines in 1972 and 1973 nested between 31 and 71% up from the bottom of the cliff, and averaged 59% (Burnham 1975). Most cliffs in the inland survey area have rounded tops due to glacial activity and few good nest ledges are found near the crown. Many ledges near the crown can be reached by fox or humans. Below the crown, most cliffs have a sheer, vertical (or near vertical) portion before adequate ledges are present. About 30% do not have good nesting ledges or locations in the upper portion of the cliff.

After conducting observations from a blind at one inland nesting site, Harris and Clement (1975) suggested that the falcons did not nest in the upper portion of the cliff because low clouds frequently reduced

Fig. 5. Peregrine falcon nest site 1 (Table 1) has a vertical exposure of rock of up to 100 m. Peregrines nested on the cliff historically and in 1972 (number 1 and arrow) and 1973 (number 2 and arrow).

visibility for hunting adults. This could be true for very high cliffs, but it certainly does not apply to low cliffs away from foggy areas.

Coastal peregrine eyries were not located on seacliffs, but were set slightly inland. We found many in protected areas in fjords or bays. Some gyrfalcons nest on cliffs near the water's edge; however, peregrines seemed to prefer sites protected from the sea.

Elevation of cliff nesting sites

During late June snow persists on the top of cliffs while the valleys are snow-free. Peregrines seemingly should select cliffs of lower, warmer elevation. But in most cases in the inland area, elevation alone does not appear to be a critical factor. Peregrine falcons nested between elevations of 100 and 500 m. The critical aspect of nesting elevation appears to be determined by a ready supply of available prey species. Cliffs examined at higher elevations, which did not overlook a lower area where small passerines were nesting, had no peregrine falcons.

Directional orientation and exposure

All but one occupied nesting cliff in the study area had southern exposures, which are substantially warmer than cliffs facing north. A pair did nest on a cliff (site 20) with a northern exposure during 1978 through 1981. Uniquely, the pair nested in an old raven *(Corvus corax)* nest. The north-facing cliffs usually appear to be much wetter and may be covered, at least in part, by the tundra-mat vegetation. The amount of vegetation and frost action is more evident on north-facing hillsides as well as cliffs.

The exposure of nesting sites averaged almost due south (Burnham 1975). Burnham determined the total hours the scrape was exposed to the sun, and when the sun first reached and later left the scrape. The results show peregrines choose nest locations which compromise between the warmth gained from the sun and the protection provided by an overhang. If the peregrine places the scrape on a projection of rock or far out on a ledge, it gains exposure to the sun but sacrifices protection and subjects its eggs and young to increased wind, rain, snow, and falling rocks (Figure 6). Falling rocks (loosened by frost action) may be an important hazard, because the cliffs are littered with glacial debris. If, at the other extreme, peregrines place their nest under the protection of a large overhang, as gyrfalcons often do, exposure to the sun is reduced. Gyrfalcons seem to have an ability to incubate and hatch eggs under lower ambient environment temperatures than peregrines. Gyrfalcons nest earlier in the spring when weather conditions are most severe and only protected ledges may be useful.

Nesting ledge

The nesting ledge is probably the single most critical factor affecting the survival of the young falcons (Figure 7). The size of an escarpment, predominance over the surrounding area, or the general directional exposure may attract falcons to a given location, but the cliff must also contain adequate nesting ledges. On cliffs with more than one good nesting location, many scrapes have been found in a single season. At one nesting site a ledge contained five deep scrapes several meters apart. The eggs were on a different ledge 30 m away.

Generally, peregrines lay their eggs on a long, narrow ledge with clumps of grass under a slight overhang of rock (Burnham 1975). In the inland area, young seem to seek shade behind grass clumps on very warm days. On a noon visit to an eyrie where vegetation was lacking, three young less than a week old showed signs of overheating until shaded by the researchers. The parent falcon apparently had been shading the young falcons before we interrupted.

Initially, we thought that peregrines sought clumps of grass when selecting a nesting location, but we now believe the grass was probably a result, and not necessarily a cause, of the falcons nesting there. The fecal matter deposited by young falcons on the nest shelf substantially increases the fertility of windblown silt. Prey remains scattered near the scrape, along with castings (an elliptical ball of undigested feathers and bone which is regurgitated), act as a compost to hold moisture and to condition the soil. These factors, combined with seeds contained in the prey transported to the nest by the adults, probably produced such dense growth. The vegetation may be desirable, but several ledges were used which virtually lack vegetation. Some factor other than vegetational protection attracted falcons to the ledge. Lack of vegetation might suggest that the ledge was being used for the first time.

Most peregrine falcon nesting scrapes examined had at least a slight overhang of rock protecting the scrape. Even though nesting scrapes at some eyries may be only partially protected from above, many offer the young protection once they are old enough to move about.

The total surface area of the nest ledge does not appear crucial in nest site selection as long as there is room for the young to lie down. The composition of the soil on the ledge does not appear to be important, as variation exists.

Peregrine density

Inland

We found a total of 34 historic and presently occupied sites in the inland area (Table 1). An "occupied site" is one where at least one peregrine falcon has been seen actively defending a cliff or displaying other nesting behavior. Cliffs checked with no peregrine occupancy one year may prove to be occupied in following years. We calculated density figures on the basis of 1972 and 1973 observations because of fluctuation in number of sites and area surveyed. In 1972, 17 known sites were checked (not all were known to have been used by peregrines when checked), 10 were occupied and 7 produced young. One of the three occupied but non-producing sites (13) was reported active by local residents in early June of 1972. At that time the falcons appeared to defend the cliff but when researchers checked the cliff in late June no falcons were present. Two other occupied but non-producing sites occurred. One had a pair of adults defending it throughout the nesting season (site 4). When researchers went to the cliff to band young in late July, a nesting scrape, castings, feathers of prey, and down from young peregrines were found. However, no young falcons were located. A hunter's camp was situated within hearing distance of the eyrie; the hunters could have disposed of the young, although we observed no signs of human (or fox) predation. At site 9 a lone female peregrine exhibited weak defensive behavior. Despite thorough observation, neither male nor young were seen. The lone peregrine had juvenile feathers remaining in her adult plumage and her age was estimated at one year. The following nine seasons a pair raised young at the same location.

In 1973, of the 17 sites checked, 9 produced young. Both of the non-productive, occupied sites (10 and 11) had been checked in 1972 and no birds were seen. Of these two sites, site 11 was located approximately 300 m from a gyrfalcon nest. A lone female peregrine defended the site for no longer than a week early in the nesting cycle (late May) and then disappeared. The gyrfalcons nested in an old raven nest and had not been seen at the cliff the year before. It is possible that the above site was an old peregrine eyrie because the cliff had suitable nesting ledges. In mid-June a pair of peregrines was observed at site 10, approximately 2 km from site 11. This pair made a nesting scrape and displayed

Fig. 6. The young peregrines, approximately 21 days old, are on a long, 1 m wide ledge which is slightly overhung by the cliff. The nest scrape where the eggs were laid is immediately in front of the young, behind the clump of grass.

early courting behavior in the form of bowing and calling. They also showed moderate aggression toward intruders. The lone female at the earlier site (11) may have found a mate late in the season at site 10. Site 10 was checked on a regular basis in 1973 because the two gyrfalcon nesting sites were being studied, each located approximately 2 km on either side of the peregrine site. The cliff was not occupied earlier that year by peregrines or gyrfalcons. Two sites occupied by gyrfalcons in 1973 were used in later years by peregrines, and a 1973 peregrine eyrie was later used by gyrfalcons.

Even when falcons did not produce young but defended a cliff, we considered the site occupied unless the falcons appeared at the site late in the breeding season. Falcons which establish a territory late probably modify nesting density insignificantly. Because of late occupancy by the pair at site 10, we calculated density for 1973 for ten sites instead of eleven (site 10 deleted), while density of producing pairs was computed on a nine site basis. Density in 1972 was computed for ten occupied sites and seven producing sites.

The size of the study area was slightly larger in 1973 than in 1972. In both years, however, about 12% of the total area included a fjord and two fingers of glacial ice extending from the inland icecap. These areas are not used for hunting by falcons or as a buffer zone between eyries, and they are therefore excluded from calculations. The tundra, containing no cliffs, and the lakes are included in density calculations as the lakes may act as buffer zones between eyries, and cliffless tundra provides hunting territory. The mean density of peregrine falcons within the 1972 and 1973 survey area (which was smaller than later years) was one occupied site per 200 km^2 and one producing pair per 244 km^2 (Burnham 1975).

These nesting densities can be compared with those reported from other areas. In 1939 in the eastern United States, Hickey (1942) found 18 pairs of peregrines and one unmated male in 25 900 km^2, or one pair/1364 km^2. Bond (1946) estimated slightly more than one pair of peregrines for every 5180 km^2 in western North America. Fyfe (1969) discussed peregrine falcon densities in Canada by ranking areas numerically to indicate habitat suitability. Fyfe's data indicate that "Group 1" areas of mainland Northwest Territories and Arctic Islands have one pair per 50 km^2. Assuming the survey area in Greenland is comparable to prime Canadian habitat ("Group 1"), then Canadian peregrines are more dense

Fig. 7. Two-day-old peregrine young at site 1 (Table 1) 3 July 1973.

in many areas. However, if "Group 2" habitat, such as the Anderson River and the Adelaide Peninsula, are compared to the Greenland area, then Greenland peregrine populations are equally dense. It is misleading to compare breeding density of inland peregrines measured as a mean area per pair with coastal breeders or river pairs where cliffs have a linear distribution.

Hickey (1969) suggested a technique developed by Ratcliffe (1972) should be used to compare peregrine density. For much of England, Wales, and southern Scotland in 1930–39, Ratcliffe (1972) determined an average value of 4.8 km between occupied coastal and inland sites, which he felt was fairly uniform. Peregrines in certain locations in England during the 1930's were apparently more dense than within the Greenland inland survey area in 1972 or 1973 where an average of 6.3 km separated eyries. The 1972–73 densities compare favorably with Ratcliffe's (1972) values for the southern Highlands and many coastal districts of the entire Scottish Highlands. The density for the southern Highlands and coastal districts was 5.5 to 6.4 km for 1930 through 1939, compared to a mean minimum distance between producing sites for the Greenland area in 1972 and 1973 of 7.7 km.

Table 2 summarizes the survey results for the study period. Even though the number of nest sites found increased over the years, the percent of occupied and successful sites remained relatively constant. The percent of known sites checked varied greatly over the years and may have caused most of the variability observed.

In a "healthy" peregrine population a certain percent of sites are unoccupied (Table 3). We found a yearly mean of 60% of the sites occupied in the inland survey over ten years. Eighty-four percent of the sites occupied by pairs produced young. The only information on an Arctic population is Cade's 1960 Colville River work where he found a 55% annual site occupancy. In Greenland, five nest sites were used by both breeding peregrines and gyrfalcons, but not simultaneously.

Coastal

More historical data are available for Greenland coastal peregrine eyries than for inland eyries. Explorers and naturalists noted locations where they found falcons nesting (Bertelsen 1932; Fencker and Scheel 1929; Oldendow 1933). In addition, they collected a number of the birds and eggs, noting nest locations or the nearest settlement. The Disko Bugt region is one of the best

Table 1. Inland peregrine nesting site occupancy.

Site number	Prior to 1972	1972	1973	1974	1975	1976	1977	1978	1979	1980	1981
1	birds present	PY(3)	PY(3)	OS	OS	OS	OS	OS	NO	NO	NO
2	U	PY(3)	PY(3)	PY(3)	PY(3)	PY(3)	PY(3)	PY(3)	PY(4)	PY(3)	PY(4)
3	U	PY(3)	PY(1)	PY(4)	PY(1)	PY(3)	OP	PY(4)	OP	PY(4)	OS
4	U	OP	PY(4)	OS	OP	OP	PY(3)	PY(4)	PY(3)	PY(4)	PY(4)
5	U	PY(1)	PY(3)	NO	NO	NO	NO	NO	NO	NO	NO
6	known locally	PY(1)	PY(3)	PY(4)	PY(1)	OG	OG	NO	NO	OS	NO
7	known locally	PY(3)	PY(2)	PY(2)	PY(3)	PY(4)	PY(1)	PY(2)	PY(3)	PY(4)	PY(2)
8	U	PY(3)	PY(2)	OP	OS	OS	NO	NO	NO	NO	NC
9	U	OS	PY(3)	PY(2)	PY(4)	PY(3)	PY(2)	PY(4)	PY(3)	PY(4)	PY(3)
10	U	NO	OP	NO	NO	NC	NC	NO	NO	NO	NO
11	U	NO	OS,OG	NO	NO	NC	NC	NO	NO	NO	NO
12	PY 1971	NO	NO	NO	NC	NC	NO	NC	NO	NC	NO
13	young taken 1940's	OP	NO	NO	NO	NO	NC	NC	NO	NC	NC
14	active for many years	NO	NO	NO	NO	NO	NO	NO	NO	NO	NC
15	reported 1970	U	NO	NO	NC	NC	NC	NC	NO	NC	NC
16	reported 1970	U	NO	NO	NC	NC	NC	NC	NO	NC	NC
17	U	U	U	OS	NC	NC	NC	NC	NC	NC	NO
18	U	U	U	OS	NC	NC	NC	NC	OS	OP	NO
19	PY(3) 1970	U	U	NO	NC	NC	NO	NC	NO	PY(2)	OP
20	U	U	U	NO,OG	NC	NC	PY(?)	PY(1)	PY(3)	PY(1)	PY(1)
21	U	OG	OG	NO	NC	NC	NC	PY(2)	OS	OP	OS
22	U	NO	OG	NC	NC	NC	NC	PY(3)	NC	NO	NC
23	U	U	U	U	U	U	U	OS?	PY(1)	PY(1)	PY(3)
24	U	U	U	U	U	PY(?)	NC	NC	PY(2)	NC	PY(?)
25	U	U	U	U	U	PY(?)	NC	NC	PY(?)	NC	NC
26	U	U	U	U	U	U	U	U	OS	NO	OP
27	U	U	U	U	U	U	U	U	OP	OS	OG
28	U	NO	NO	NC	NC	NC	NC	NC	PY(2)	NO	PY(3)
29	U	U	U	U	U	U	U	U	U	PY(2)	PY(4)
30	U	U	U	U	U	U	U	U	U	PY(4)	PY(3)
31	U	U	U	U	U	U	U	U	U	PY(4)	PY(3)
32	U	U	U	U	U	U	U	U	U	PY(3)	PY(3)
33	U	U	U	U	U	U	U	U	U	U	OS
34	U	U	U	U	U	U	U	U	U	U	PY(2)

U = unknown. NC = not checked. NO = not occupied. OS = occupied by a single adult peregrine. ( ) = number of young produced. PY = produced young. OG = occupied by gyrfalcons. OP = occupied by pair of peregrines.

Note: Some sites listed on Table 1 are outside of the survey area proper and therefore variation exists between Table 1 and Table 2.

known areas with eight historical nesting sites (Table 4) and numerous sightings.

From the available literature, field research, and reported sightings, the coastal population of peregrines in Greenland appears sparsely scattered. Most reported eyries are found in fjords or in protected locations. The sites seem to be situated near concentrations of prey. For example, sites 1 and 2 are close to an alluvial flood plain which has numbers of purple sandpipers *(Calidris maritima)* and ringed plovers *(Charadrius hiaticula)*. We found numerous references in the literature which suggest that Greenland peregrines feed largely on seabirds in coastal areas; however, site locations and prey remains collected from the one active eyrie suggest that they feed mostly on passerines.

Temperature and weather conditions along most of Greenland's west coast are fairly uniform. The mean annual temperature difference between Upernavik (73°N) and Ivigtut (62°N) is 10C°. Precipitation increases southward along the coast. The climate conditions along the coast are similar enough so that information developed in one coastal location may reflect the peregrine's status in other coastal areas.

W. Burnham and S. Sherrod examined 395 linear km of coastline in the Disko Bugt area in 1974. They examined five confirmed peregrine nesting sites within the area. One of the five sites was occupied. That site (7, Table 4) had a juvenile killed in the area in August of 1879 (Schiøler 1925–31). If all sites were occupied, the mean minimum distance between sites would be 55 km. Of the five sites, only one was occupied and it produced four young. Two sites not occupied in 1974 were reported by H. Fencker (pers. comm.) to have been used in 1973. One of these is an historical site (1 or 2, Table 4) reported in 1948.

In 1979 W. Burnham and W. Burnham re-examined two of the 1974 sites. The same site (7) was again occupied and young were present. The second, known since 1859, is near Godhavn and was not occupied. Greenlanders were camping below the cliff. The site

Table 2. Peregrine falcon. Summary of effort, occupancy and reproduction success 1972–1981.

Year	Effort			Occupancy				Reproduction success						
	Known cliffs	No. checked	% of known	Un-occ.	Lone adult	Pairs	% occ.	Pairs w.young	% of pairs	Total young	Young/ occ. site	Young/ pair	Young/ succ. pair	Young/ cliff check.
1972	17	17	100	7*	1	9	59	7	78	17	1.70	1.88	2.42	1.00
1973	17	17	100	6**	1	10	65	9	90	24	2.18	2.40	2.66	1.41
1974	21	19	90	9	4	6	53	5	83	15	1.50	2.50	3.00	0.79
1975	21	13	62	5	2	6	62	5	83	12	1.50	2.00	2.40	0.92
1976	21	11	52	4*	2	5	64	4	80	13	1.86	2.60	3.25	1.18
1977	21	12	57	6*	1	5	50	4	80	9	1.50	1.80	2.25	0.75
1978	22	16	73	6	2	8	62	8	100	23	2.30	2.88	2.88	1.44
1979	24	22	92	10	3	9	55	7	78	19	1.58	2.11	2.71	0.86
1980	28	25	89	9	2	14	64	12	86	36	2.25	2.57	3.00	1.44
1981	30	26	87	9*	3	14	65	12	86	35	2.06	2.50	2.92	1.35

* 1 cliff occupied by gyrfalcons
** 2 cliffs occupied by gyrfalcons

Table 3. Peregrine falcon. Percent occupancy of nesting sites.

Location	Time period	No. sites	Percent occ./year	Reference
Colville River, Alaska	mid 1950's		55%	(Cade 1960)
Mississippi River drainage	1954–55	14	71%	(Berger & Mueller 1969)
Massachusetts	1935–42	14	77%	(Hagar 1969)
Hudson River	1950–54	8	80%	(Herbert & Herbert 1969)
Pennsylvania & New Jersey	1939–41	12–20	85%	(Rice 1969)
Great Britain	1930–39	805	85%	(Ratcliffe 1972)
West Greenland	1972–81	14–34	60%	

with young, unlike the second, is on a high cliff composed of loose rock, over 100 m above sea level and 1 km inland.

It is difficult to explain the low occupancy of the coastal sites examined. The Greenlanders probably have a substantial impact on the birds, and those near settlements or within boat range could be easily disturbed. Gulls which nest within boathook reach of the water are eaten; low nesting falcons could be caught or shot for human consumption. Many areas searched had a limited number of cliffs and all were checked. Searches in two summers provided little insight into the situation.

Table 4. Coastal peregrine nesting sites – Disko Bugt.

Site No.	History	1974	1979
1[ad]	1948 juvenile collected, reported active 1973	NO	NC
2[ad]	1973 reported active	NO	NC
3[d]	1880 eggs taken and one adult shot, 1948 juvenile collected	NC	NC
4[bc]	1859 adult female collected, 1877 both adults collected	NO	NO
5[c]	1880's active	NC	NC
6[c]	1880's active	NO	NC
7[bc]	1879 immature female collected	PY(4)	PY(?)
8[e]	1951 dead fledgling found under cliff	NC	NC

[a]. Site 1 and 2 are very close to each other and the juvenile female could have come from either location. [b]. (Schiøler 1925–31). [c]. (Fencker and Scheel 1929). [d]. Universitetets Zoologiske Museum. [e]. Mattox.

NO = not occupied. NC = not checked. PY = produced young. ( ) = number of young present.

Peregrine productivity

Productivity is the average number of young per nesting pair a given population fledges into the wild. The number of young which fledge from eyries is usually overestimated because older nestlings last seen are assumed to have later fledged. Nestling mortality appears to be high at two times. The first is from hatching to three days of age, and the second appears to be prior to fledging when young compete for food, causing younger or slower eyasses to die (Figure 8). Mortality is also high before fledged young reach independence and disperse (Burnham, unpubl. data), but this is not considered in the production rate.

Production data for all sites occupied by pairs (Table 2) in the inland survey area over the ten year period exceeds Hickey and Anderson's (1969) estimated 1.16 young per year required from all pairs of breeding peregrines to maintain a stable population. Mean production for Greenland birds is 1.8 young per occupied site

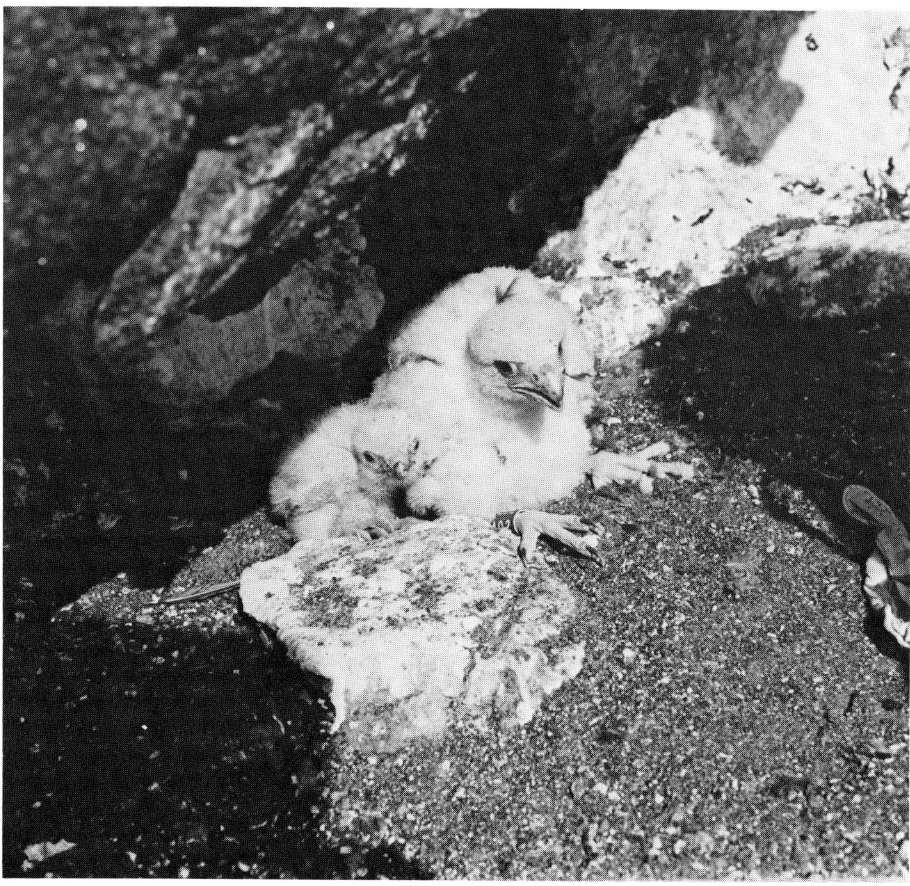

Fig. 8. The oldest of these two peregrines, a female, is approximately 17 days old. The smaller, a male, is approximately 10 days old. When the site (6, Table 1) was revisited three days later, the smaller young was gone. Odd aged young have been observed at eyries in Greenland several times but none have survived to flying.

(pairs or single adult), 2.3 young per site occupied by a pair, and 2.8 young per successful pair. The number of young reared per successful pair in England (Ryves 1948) and North America (Hickey 1942) was 2.4 and 2.5 respectively prior to introduction of DDT. In these terms, the peregrine falcon population in the study area appears to be healthy and thriving (Table 5).

Peregrine prey

Prey species

Table 6 lists all birds observed within the inland study area. The list includes birds eaten by nesting peregrines, unusual species, and species rare to the area. Only four small passerines occur regularly in the area: Lapland longspur, snow bunting, wheatear, and common redpoll. All four are summer residents arriving during May. After nesting they depart between the last two weeks of August and mid-October (Salomonsen 1967).

Prey remains were collected from 14 different active peregrine eyries (13 inland, 1 coastal) in 1973. The prey remains mostly represent those species captured to feed young, but one pair did not lay eggs; prey remains collected there corresponded to those taken by other pairs feeding young. All four of the small passerines listed above are preyed upon by the peregrine falcons in our inland study area during the summer and make up over 90% of the peregrine's diet in total number and biomass (Table 7) (Figure 9).

Approximately 70% of the inland peregrine falcon diet consisted of Lapland longspur in 1973 (Table 7). At the coastal eyrie a similar part of the peregrine diet consisted of snow buntings. Small bird censuses were made, and those data, combined with prey remains found in eyries, suggest that generally the peregrines capture more individuals of the most abundant species. Other prey species found include northern phalarope, oldsquaw, rock ptarmigan, and arctic hare. Hare and ptarmigan, when available, make up most of the gyrfalcon's diet.

Prey density

W. Burnham conducted small bird transects throughout much of the survey area between 22 May and 27 July 1973. A transect consisted of an area 900 m long and 30 m wide. It was made by walking 1000 steps in a nearly straight line and counting all birds observed or heard

Table 5. Annual peregrine brood size.

Year	Young/nest				Known sex of eyasses	
	4 young /site	3 young /site	2 young /site	1 young /site	M	F
1972	1	4	–	2	9	4
1973	1	5	2	1	19	5
1974	2	3	–	1	13	5
1975	1	2	–	2	6	6
1976	1	3	–	–	3	7
1977	–	2	1	1	3	6
1978	3	2	2	1	7	10
1979	1	4	2	2	9	11
1980	6	2	2	2	16	14
1981	3	6	2	1	16	16
Total	19 (25%)	33 (44%)	11 (15%)	13 (17%)	101 (55%)	84 (45%)

vocalizing within 15 m on either side. Certain areas were measured and marked, then periodically walked as standards. This technique was used by C. White, T. Cade, and J. Haugh (1967 unpubl. data) on the Colville River in Alaska. In Greenland we found an average of 5.8 passerines per transect, which is much below the value (average 32 birds per transect) found by similar studies in Alaska (C. White, T. Cade, J. Haugh 1967, unpubl.). In addition, a much greater diversity of species occurs in Alaska. We have no data on yearly variation. Transects near two abandoned eyries (peregrines reported in 1969 and 1970, N. Ten Brink, pers. comm.) gave an average below 1.0. The absence of falcons on apparently suitable cliffs may be an indication that the extremely low prey density is at least one factor affecting occupancy. If this were true, however, it is curious that falcons ever nested in the area, unless prey or peregrine density has decreased.

Table 6. Birds seen in the inland study area.

Common name	Scientific name	Abundance
Common loon	*Gavia immer*	common*
Arctic loon	*Gavia arctica*	accidental
Red-throated loon	*Gavia stellata*	common*
Great cormorant	*Phalacrocorax carbo*	common* (1 colony in study area)
Canada goose	*Branta canadensis*	rare (1 group of 3 seen)
Brant	*Branta bernicla*	uncommon (1 flock of 12 seen)
White-fronted goose	*Anser albifrons*	common*
Mallard	*Anas platyrhynchos*	common*
Oldsquaw	*Clangula hyemalis*	common**
Red-breasted merganser	*Mergus serrator*	common*
White-tailed sea eagle	*Haliaeetus albicilla*	uncommon****
Gyrfalcon	*Falco rusticolus*	common*
Peregrine falcon	*Falco peregrinus*	common*
Rock ptarmigan	*Lagopus mutus*	common**
Grey heron	*Ardea cinerea*	accidental
Ringed plover	*Charadrius hiaticula*	uncommon*
Purple sandpiper	*Calidris maritima*	common locally*
Northern phalarope	*Phalaropus lobatus*	common**
Ivory gull	*Pagophila eburnea*	rare
Glaucous gull	*Larus hyperboreus*	common*
Iceland gull	*Larus glaucoides*	common*
Herring gull	*Larus argentatus*	rare
Black-headed gull	*Larus ridibundus*	uncommon***
Black-legged kittiwake	*Rissa tridactyla*	uncommon*
Northern raven	*Corvus corax*	common*
Wheatear	*Oenanthe oenanthe*	common**
Common redpoll	*Carduelis flammea*	common**
Lapland longspur	*Calcarius lapponicus*	common**
Snow bunting	*Plectrophenax nivalis*	common**

* Species which breed in study area.
** Species which breed in study area and are preyed upon by peregrine falcons.
*** At least one and usually two individuals of this species were seen regularly at a gull colony on the fjord during 1972 and 1973. It is not known if they nest in the area.
**** Both immature and mature eagles have been seen within the study area. Three large stick nests were found by aircraft on a section of cliff more than 120 m high. However, time did not allow the checking of these nests to determine if the eagles nested there during the study period.

Gyrfalcon biology

Cade (1960) estimated the population of gyrfalcons at over 1000 breeding pairs in Greenland. Native Greenlanders are quite aware of the presence of this species, much more so than of the peregrine, and know many nest locations in coastal areas. The large inland study area near Søndre Strømfjord is less familiar to the Greenlander because of vast areas inaccessible by navigable waters.

Inland

Within the inland area we found gyrfalcons nesting at 22 locations during the ten-year period (Table 8). Considering that 12 of the occupied peregrine eyries did not produce young during the study period, and that we did not arrive early enough each spring to observe gyrfalcons courting and establishing territories, we believe that the numbers of gyrfalcon and peregrine nesting sites are similar in the inland area.

At many sites gyrfalcons depend on raven nests (Figures 10 and 11). The survey area surrounds Søndrestrømfjord Air Base and the Danish facility which foster a much greater density of ravens than any other location we have observed in Greenland. Ravens are shot in most of Greenland, but not near the Air Force base. The ravens feed largely at the base's refuse dump. We found castings with foil, plastic, and glass imbedded in them at raven nests over 40 km from the dump. The

Fig. 9. The adult female peregrine prepares to feed a partially plucked passerine to the 21-day-old young.

Table 7. Prey species and portion taken by peregrine falcon, 1973.

Nest attempt	Lapland longspur	Snow bunting	Wheatear	Common redpoll	Northern phalarope	Rock ptarmigan	Oldsquaw	Arctic hare
1	.7	.3	Tr	Tr	Tr			
2	.6	.3	.1			Tr	Tr	Tr
3	.9	.1				Tr		
4	.8	.2						
5	.6	.3		Tr	.1			
6	.9	.1			Tr			
7	.8	.2						
8	.8	.1				Tr		
9	.5	.2			.2	.1		
10	.5	.5	Tr		Tr			
11	.9	.1	Tr		Tr			
12	.9	Tr			Tr			
13	.9	.1						
Mean	.7	.2	Tr	Tr	.1	Tr	Tr	Tr

Tr = trace (those species taken less than .1 of the time). 1.0 = total diet of the peregrine during nesting season.

large number of ravens may influence the local gyrfalcon population by providing an increased number of available stick nests. About 65% of the inland gyrfalcon eyries are old raven nests (Table 8). A raven nest used by gyrfalcons will usually be torn apart by the young during the pre-fledging period. Before the gyrfalcons can reoccupy the site, therefore, ravens must at least attempt breeding and rebuild the nest or build a nest elsewhere on the cliff the same year. The sites where pairs do not depend upon raven nests are occupied more frequently than cliffs with no permanent eyrie location. We found one gyrfalcon eyrie with a layer of excrement over 2 m thick, all of it appearing to be from gyrfalcons.

Many nesting sites used by gyrfalcons that appeared to be unoccupied on first examination were later determined to be occupied by one or both adults. The falcons occupied sites even though not attempting to nest, especially at locations where availability of stick nests was not a factor. The pairs dependent on stick nests were perhaps more nomadic or maintained a larger territory which may include more than one section of cliff. The distance between nesting sites based on 1973 and 1974 data is 10.4 km for occupied sites and 12.3 km for successful sites.

The inland gyrfalcons feed on a variety of food items (Jenkins 1978). We found passerine feathers in abundance in gyrfalcon eyries, especially during years when

Table 8. Inland gyrfalcon nesting site occupancy.

Site number	Prior to 1972	1972	1973	1974	1975	1976	1977	1978	1979	1980	1981
1	U	PY(3)[cd]	NO	NO	NO	NO	NO	NC	NO	NC	NC
2	U	PY(2)[bd]	NC	NC	NC	NC	NC	NC	NC	NC	NC
3	U	PY(3)[bd]	PY(2)[cf]	NO	NC	NC	NC	P	P	P	P
4	U	U	PY(3)[cd]	NC	NC	NC	NC	PY(?)	PY(?)	PY(?)	PY(?)
5	U	PY(?)[ad]	PY(3)[ad]	O[e]	NC	NC	NC	NC	PY(2)[ad]	NC	PY(2)[d]
6	U	U	O(?)[h]	NC	NC	NC	NC	NC	PY(1+)[d]	NC	NC
7	known by natives	O[h]	PY(2)[cg]	O[h]	NO	NO	NO	PY(?)	NC	NC	NC
8	U	O[e]	NC	NC	NC	NC	NC	P	NC	NO	NC
9	known by natives	PY(?)[ad]	O[e]	PY(4)[af]	PY(2)[af]	O	NC	PY(?)	NC	NC	NC
10	U	U	NO	PY(3)[cf]	NC	NC	NC	NC	NC	NO	NC
11	U	U	U	PY(?)[a]	NC	NC	P	P	P	P	P
12	U	U	U	PY(3)[cd]	NC	NC	PY(2)	PY(3)[h]	NC	NO	NO
13	known by natives as a peregrine eyrie for many years	P	P	P	P	PY[cg]	PY[cg]	NO	NO[i]	P	NO
14	U	U	U	U	U	O	NC	NC	NC	NC	NC
15	U	U	U	U	U	U	U	U	PY(1+)	NC	NC
16	U	U	U	U	U	U	U	U	U	PY(1+)	NC
17	U	U	U	U	U	U	U	U	U	O[h]	NC
18	U	U	U	U	U	U	U	U	U	O[e]	NO
19	U	U	U	U	U	U	U	U	U	NO	PY(3)
20	U	U	U	U	U	U	U	U	U	U	PY(1+)[f]
21	U	U	U	U	U	U	U	U	U	U	O[e]
22	U	U	U	U	U	U	U	U	P	P	PY(?)

a. Permanent nest composition. b. Nest structure unknown. c. Raven nest used. d. White pair. e. One white bird seen or molted feathers found. f. One white bird and one grey bird. g. Grey pair. h. One grey bird seen or molted feathers found. i. Cliff fallen away.

U = unknown. NC = not checked. P = occupied by peregrines. NO = not occupied. PY = produced young. O = occupied. ( ) = number of young.

hare and ptarmigan numbers were low. At no time during our ten-year study did the ptarmigan population average more than one ptarmigan seen per day. The duck population is also limited, and during the period of maximum food requirement of young gyrfalcons, the ducks are nesting with many flightless birds remaining on ponds and lakes where they are not vulnerable to attack. The gyrfalcon is partly dependent on passerines and arctic hares for food. The energy expenditure in catching a small passerine compared to energy gain when consumed by an adult or fed to young must be very great. However, in the low vegetation young passerines are probably easy to catch so the hunting success may be high. The prey remains and castings at one eyrie site from which four young fledged consisted almost totally of small bird feathers. The eyrie was located in an area of above average passerine density.

Coastal

We found, in various sources, 18 gyrfalcon eyries reported from the Disko Bugt study area. Of these, we examined 14 sites in 1974, but no young were found. We saw adults near some locations and freshly molted gyrfalcon feathers at other sites. Most of the locations had definitely been used by gyrfalcons in the past. Local Greenlanders in the area said the eyries were not usually used each year. The ptarmigan and hare numbers on the coast were very low in 1974. In general, we saw fewer hares in coastal areas than inland. In the Disko Bugt area large numbers of gulls and seabirds may provide alternate quarry for gyrfalcons. There are reports in the literature of gyrfalcons feeding on such prey, but we found no active sites near seabird colonies. The reason no producing sites were found was not a lack of field effort; we spent much time climbing near reported and potential eyries.

The gyrfalcon throughout the North American Arctic shows no eggshell thinning and the population appears stable. The absence of breeding gyrfalcons in 1974 may have been a response to low ptarmigan and hare numbers. Similar periodic fluctuations have been observed in the Alaskan Arctic.

The mean minimum distance between reported gyrfalcon nesting sites for the coastal area was 15 km. The peregrine site to gyrfalcon site ratio in the Disko area is probably about 1 to 3. This ratio could be representative of other coastal areas south of Disko Bugt. North in the Upernavik district a further reduction in the number of peregrine nests is reported (Bertelsen 1932). Limited research and the available literature on Greenland's east coast falcon populations indicate a further exaggeration in the ratio because of fewer peregrines.

Fig. 10. Gyrfalcons nested (arrow) in an old raven nest (Fig. 11), 51 m down from the top edge of the cliff (site 7, Table 8).

Interspecific competition and falcon density

Three raptors nest in the study area: the peregrine falcon, gyrfalcon, and the white-tailed sea eagle *(Haliaeetus albicilla)*. The raven, another large, cliff-nesting bird which occupies the area, has been referred to as a functional raptor by White and Cade (1971) because of its occasional predatory role.

Each of the birds listed above may influence the nesting density of the others. White-tailed sea eagles are not numerous anywhere in Greenland so their effect on falcon density, as a result of occupying cliff sites, appears to be negligible. Few apparently "good" nest cliffs in the survey area were unoccupied by falcons where adequate prey density (5.8 birds/transit) exists at least periodically. Ravens nest on smaller and/or apparently less desirable cliffs, as well as on the same cliffs as both gyrfalcons and peregrines. Ravens appear to have little direct effect on peregrine density. The raven's diet is quite broad-based and offers little potential competition. Since the gyrfalcon is partially dependent on ravens for nest construction, the raven directly affects gyrfalcon nesting. Ravens may then indirectly impact peregrines since we have not found peregrines and gyrfalcons breeding simultaneously on a cliff in Greenland. The two species did occupy the same cliff briefly in 1973 when a lone female peregrine defended a section of cliff on which gyrfalcons nested (gyrfalcon site 4, Table 8, and peregrine site 11, Table 1).

Nesting by gyrfalcons and ravens begins about the same time while peregrines breed later in the spring. Because of the gyrfalcon's larger size and power, it can displace either ravens or peregrines. W. Burnham observed an adult male peregrine diving at and chasing an adult white male gyrfalcon. When the peregrine dropped below the gyrfalcon and was at a disadvantage, the gyrfalcon began chasing the peregrine. When they were almost out of sight through a 20 power scope, the two birds merged and did not separate. The peregrine may have been killed by the gyrfalcon, but we have not found peregrine remains at gyrfalcon eyries.

Fig. 11. Two young gyrfalcons are shown in an old raven nest (site 7, Table 8). This nest is indicated by the arrow in Fig. 10. Notice molted primary feather from gray female parent (arrow, left) and front leg of caribou (arrow, right) apparently left by ravens.

On the Colville River in northern Alaska, gyrfalcons and peregrines have frequently been found nesting on the same cliff as close as 35 m (White and Cade 1971). This same area supports a wide variety of small birds (five times more than occurred in inland Greenland), ptarmigan, shore birds, water birds, and microtine rodents. Ptarmigan in Alaska make up over 90% of the gyrfalcon's diet but are seldom found in peregrine eyries, so little overlap in diet occurs. In Greenland in the inland study area where fewer prey species and individuals of each exist compared to Alaska, a far greater degree of prey species overlap occurs between the peregrine and gyrfalcon, especially in years of few ptarmigan and hares. When prey is limited in Greenland and where a great degree of diet overlap occurs, competitive behavior may have evolved and explains why the two species do not simultaneously nest on a cliff. Under such competition the peregrine would be displaced. However, the gyrfalcon's greater dependence on larger, more widely scattered prey may act to disperse gyrfalcon pairs. That, combined with the gyrfalcon's partial dependence on raven nests, may reduce the gyrfalcon's potential dominance and provide a greater opportunity for peregrine nesting and reduce interspecific competition.

If the above analysis is correct and competition exists for nest sites or breeding areas, then what is the density-limiting factor for peregrines? We believe that peregrines produce adequate numbers of young for recruitment although post-fledging mortality may be high and is unknown. We know that gyrfalcons are partially dependent on raven nests and an adequate food supply to attempt breeding. We also believe that gyrfalcons can displace peregrines. By our observations we determined that most larger cliffs at proper elevations near adequate prey density in the inland survey area have been occupied by falcons. It would seem, then, that since gyrfalcons and peregrines have not been found breeding simultaneously on a cliff, gyrfalcons, along with a constant number of cliffs, directly affect peregrine density. Indirectly, however, prey abundance is probably the single greatest factor which affects peregrine density in the inland survey area. If "unlimited" prey availability existed then we would see a situation similar to Alaska where gyrfalcons and peregrines nest on the same cliff.

Pollutants and their effect

The one potential threat to the peregrine falcon's existence in Greenland is contamination with DDT and, possibly, polychlorinated biphenyls (PCB). Even though we found no population decline or decrease in production of young during the ten-year study, substantial eggshell thinning and moderate DDE, a metabolite of DDT, and PCB levels occur in eggs. The mean DDE residue level of 14.3 ppm (wet weight) (Table 9) for all peregrine eggs analyzed during the study is lower than levels reported for other arctic and sub-arctic peregrines.

Cade et al. (1971) reported a mean of about 42 ppm for 19 peregrine eggs collected on the Colville River in Alaska from 1967 to 1969. They also report DDE levels of about 32 ppm for eggs collected in the Yukon and Canada in 1968 and 1969. Burnham et al. (1978) reported a mean DDE level of about 35 ppm for five Colorado peregrine eggs. DDE levels of about 7.8 ppm were reported for eleven eggs from Amchitka, Alaska, peregrines during 1969 and 1970 (Cade et al. 1971). Eggs from Amchitka peregrines contain high PCB levels, about 32 ppm for three eggs collected in 1970 (White and Risebrough 1977). The mean of 17.1 ppm PCB for the eight Greenland peregrine eggs is three times higher than levels reported for Colorado peregrine eggs; however, the Greenland mean is only half the level reported for Amchitka peregrine eggs (White and Risebrough 1977; Enderson and Craig 1974). PCB-associated behavior abnormalities, embryonic death, and eggshell thinning have been reported in some bird species (Hays and Risebrough 1972; Peakall and Peakall 1973).

Table 9. DDE and PCB levels in addled peregrine falcon eggs.

Year	Site	ppm DDE Lipid	ppm DDE Wet[1]	ppm PCB Lipid	ppm PCB Wet[1]
1972[2]	6	364	17.1	403	18.9
1972[2,3]	8	300	14.1	210	9.9
1973	8	364	17.1	148	6.9
1977	2	227	10.6	647	30.4
1977	4	94	4.4	119	5.6
1977	9	211	10.0	285	13.5
1978	7	392	18.5	580	27.4
1978	20	488	22.8	190	8.9
Mean (arithmetic)		305	14.3	323	17.1
(logarithmic)		275	13.0	278	13.2

1. Wet wt. values estimated for 1972–73 on basis of a wet weight: lipid wet ratio of 21.3 for 1977–78 eggs. 2. (Walker et al. 1973). 3. Values for an embryo.

Forty-two peregrine eggshells collected from Greenland prior to 1940 were measured and averaged 0.347 mm ± 0.018 in thickness (D. W. Anderson, unpubl. data). Pre-1947 peregrine eggs from Alberta, Saskatchewan, and Montana had a 0.359 mm ± 0.005 mean thickness (Anderson and Hickey 1972). Eastern United States eggs collected pre-1947 were even thicker with a mean of 0.375 mm ± 0.005 (Anderson and Hickey 1972). The mean thickness of Greenland peregrine eggshells (1972–78) is 0.291 mm (with membrane) or 16% thinner than pre-1940 eggs (Table 10). We have detected no trends in shell thicknesses through the study.

The 16% mean thinning for the Greenland peregrine eggs is less than the 21% observed for Colorado eggs in 1976 (Burnham et al. 1978). If percent thinning for Ungava eggs collected in 1967 and 1970 is calculated using mean thickness for pre-1940 Greenland eggs rather than the Saskatchewan or eastern United States eggs, the level of thinning is identical to recent Greenland eggs (Berger et al. 1970).

Using eggs we collected in 1972, Walker et al. (1973) placed the Greenland population at a contamination level near critical and suggested that a small increase in DDE levels would endanger the population. We have determined no apparent changes in levels of contamination of eggs or change in shell thickness between samples collected early and late in the study period. We believe, therefore, that the early conclusion by Walker et al. continues to be true.

In 1973 and 1979 we collected representative samples of Lapland longspur, snow bunting, redpoll, wheatear, and northern phalarope for tissue analysis for DDE and PCB (Table 11). Birds were analyzed as individuals, the results separated by year and area collected, then means determined for each species. All birds collected carried low levels of DDE and PCB. The Lapland longspur carried the highest level of DDE, but well below the level to induce shell thinning if ingested. The longspur was the species most frequently fed on by peregrines. From these results we believe that the levels of DDE found in peregrine eggs and the correlated degree of shell thinning does not result from feeding on prey in the nesting area.

Migration and banding recoveries

Peregrine falcon

Prior to 1972 little was known about migration of Greenland peregrines. Two records of birds banded in Greenland and recovered elsewhere had been received (Salomonsen 1967): a nestling banded by R. Meredith

Table 10. Peregrine falcon eggshell thickness by site and year (with membrane mm/without membrane mm).

Site	1972	1973	1974	1977	1978	1979	Mean
1	.290/.218*	.310/.250*					.300/.234
2		.299/.239		.304/.244		.288/.228**	.302/.242
3	.260/.200					.287/.227**	.260/.200
4				.288/.236*		.279/.219**	.288/.236
5	.310/.220*						.310/.220
6		.300/.240	.273/.224				.287/.232
7		.278/.215		.273/.213**	.306/.246	.250/.190**	.287/.230
8	.310/.250	.325/.246					.318/.248
9		.282/.222	.284/.221	.307/.230			.291/.224
20					.269/.205		.269/.205
28						.263/.203**	
Mean	.287/.223	.299/.235	.279/.223	.300/.237	.288/.226		.291/.229

* More than one egg in sample. ** Poor sample, omitted from means.

near Narssarssuaq in West Greenland on 4 August 1941 was recovered near Cienfuegos, Cuba, on 2 December 1941. And a nestling banded in Umanak district, West Greenland in July, 1956, was recovered on 16 October 1956 at Windigo River, Quebec (about 150 km north of Montreal). Two peregrine falcons banded in October 1956 and 1957 at Assateague Island, Maryland, were recovered (shot) later in West Greenland. The falcon banded in 1956 (October 5) was recovered in November, 1959, near Kangatsiaq, West Greenland; the falcon banded on 10 October 1957 was found shot in Umanak district, West Greenland, on 4 September 1958. These few returns suggest that Greenland peregrines migrate past eastern Canada along the east coast of the United States, south at least as far as Cuba.

One of the earliest references to the migration of peregrines nesting in Greenland was made by Winge (1898) who stated:

> The peregrine is not uncommon as a breeding bird along Greenland's west coast, both in south and north. As a migratory bird it . . . arrives comparatively late to the island, and it migrates away again in October. On migration over the seas around Greenland it has been seen many times. Over Davis Strait near the west coast it was seen by Sabine in the third week of September 1818 . . . Holbøll has seen it two times over the Atlantic Ocean southwest of Iceland. (auth.transl.)

Schiøler (1925–31), p. 404, contributed more detail about migration dates:

> The peregrine migrates away from Greenland in the winter . . . the migration of immature birds in autumn occurs especially in the month of September after some few birds begin the migration at the end of August; after that the migration of young birds extends into October until about 15–17 October or a bit later; as late as 7 to 13 November two immature birds have been shot. The adult birds arrive in the middle of May (which was reported by Holbøll already in 1846), and their migration to the south appears to fall somewhat earlier than the immature birds, because after 24 September no adult birds have been shot in Greenland. (auth.transl.)

Table 11. DDE and PCB levels in tissues of certain falcon prey species*.

		DDE ppm			PCB ppm		
		Wet wt.		Lipid wt.	Wet wt.		Lipid wt.
Species	Number	Mean	Range	Approx. mean	Mean	Range	Approx. mean
West Greenland							
Inland survey area:							
1973** Lapland longspur	5	0.43	0.02–1.70	10.3			
Snow bunting	1	0.08		1.9			
Northern phalarope	3	0.06	0.02–0.10	1.4			
1979 Lapland longspur	8	0.38	0.10–0.94	6.7	0.027	0.045–0.016	0.51
Lapland longspur (juv.)	1	0.043		1.3	<0.010		
Snow bunting	4	0.08	0.032–0.11	1.64	0.02	0.033–0.018	0.47
Northern phalarope	1	0.73		4.6	0.036		0.23
Redpoll	6	0.059	0.017–0.15	1.32	<0.013		
Redpoll (juv.)	5	0.033	0.013–0.067	0.58	<0.013		
Wheatear (juv. & adult)	7	0.123	0.070–0.17	2.3	0.026	0.038–0.016	0.49
East Greenland**							
Coastal:							
1973 Lapland longspur	1	0.04		1.0			
Snow bunting	2	0.07	0.06–0.07	1.7			
Wheatear	2	0.18	0.12–0.23	4.3			

* Whole body analysis. ** (Burnham 1975)

Oldendow (1933), however, reports observations of peregrines wintering in Greenland:

It is certainly a migratory bird on the west coast . . . but this is not completely the case because at any rate in certain winters it can be seen in Godthaab District all the time. (auth.transl.)

Oldendow, who had never personally seen or shot a peregrine in the middle of winter in Greenland, had observed them as late as 17 November (1928) and as early as 26 March (1926), both times at the Kookøerne off Godthaab. On 8 February 1932, while anchored at Kangeq, Oldendow received a peregrine shot there that day by a Greenlandic hunter, and another hunter shot a peregrine between Christmas and New Year 1931/32 at the same place (Oldendow 1933). Oldendow suggests it is possible that "these winter occurrences are connected with the recent, quite unusually mild Greenlandic winters which have resulted in a part of the species' most favored prey, the smaller birds, wintering up here in larger and larger number" (auth. transl.).

Salomonsen (1967) reports that the autumn migration of peregrines from Greenland "takes place from the end of August to the end of October, seldom later, in that the main migration takes place in September" (auth. transl.).

Other than Meredith's isolated banding in 1941, only 35 peregrines had been banded in Greenland from 1946 to 1972. In the 10 years since 1972, our survey has banded 185 peregrines, eight of which have been recovered (4.3%). Five of these recoveries support an eastern North American autumn migratory route and wintering in South America (see Table 12 and Figure 12).

During the autumns of 1975 and 1976 W. Mattox carried out banding attempts on the southwest coast near Frederikshåb. In 1975 he manned a trapping site with D. Berger and W. Clark from 7 September until 20 September, during which period they saw several peregrine falcons, and trapped one, an immature male, on 18 September which was banded and released. The first peregrine sighted was on 12 September. In 1976 W. Mattox and S. Sherrod manned a trap site 10 km to the northwest of the 1975 site from 28 September to 3 October, during which period they observed only one peregrine (immature female). The peregrine was chased off the lure bird by an immature male gyrfalcon (28 September). A storm on 1–2 October with north winds of 30–40 mph made further trapping impossible.

S. Sherrod placed radio transmitters on fledgling peregrines at two sites near Søndre Strømfjord in 1976 to study post-fledging behavior. The latest date which peregrines remained in the survey area as indicated by radio contact was 19 September (S. Sherrod pers. comm.).

Passage routes from Greenland southward are unclear, although band recoveries noted above suggest a route across Davis Strait, Baffin Island, and Labrador, thence inland, perhaps to the east shore of Hudson Bay and to the Atlantic coast of the United States.

Another possible migration path is from the southern tip of Greenland to Newfoundland, the Canadian Maritime provinces, along the coast of New England, or direct to the outer barrier islands of Maryland-Virginia-Carolinas, USA. An oceanic route from North America has recently been described for small birds detected by radar. Perhaps peregrines on their way from Greenland to South America also follow this path. A sighting (recovery no. 5) of a Greenland banded bird by D. Holcomb on the Russian research vessel Belagorsk on 3 October 1979, 200 miles east of Montauk Point (F. P. Ward unpubl. ms.) supports this migration route. Holcomb was conducting a census of seabirds for Manomet Bird Observatory and may have intercepted an oceanic migration. Such a migration phenomenon has been described recently (Williams and Williams 1978), whereby radar observations have detected at

Table 12. Recoveries of peregrine falcons banded in Greenland, 1972–1981.

No.	Date banded	Site	Sex	Date recovered	Location	Lat.-Long.	Remarks	Reference
1	27 July 1974	7	M	13 October 1974	Cape Charles, VA, USA	37°10′N–76°00′W	trapped & released	Clark, W.
2	29 July 1974	9	M	12 June 1975	Egedesminde District (West Greenland)	68°03′N–52°40′W	found dead	Zoologisk Mus.
3	2 August 1974	3	M	21 December 1975	Guayas, Ecuador	02°25′S–80°22′W	shot	Zoologisk Mus.
4	28 July 1978	3	F	12 October 1978	Cape May, NJ, USA	38°50′N–74°50′W	trapped & released	Clark, W.
5	26 July 1979	2	M	3–4 October 1979	at sea	40°26′N–65°40′W	red band no. read	Holcomb, D.
6	?	?	?	7 October 1979	Cape Sable, NS, Canada	43°05′N–65°40′W	red band sighted	Willis, P.
7	23 July 1980	3	M	26 September 1980	Democrat Pt., NY, USA	40°37′N–73°18′W	trapped & released	Safina, C.
8	5 August 1981	2	F	12 October 1981	Fisherman's I., VA, USA	37°10′N–76°00′W	trapped & released	Bird, M.

Total banded = 185. Total shot = 1. Total captured and released = 4.

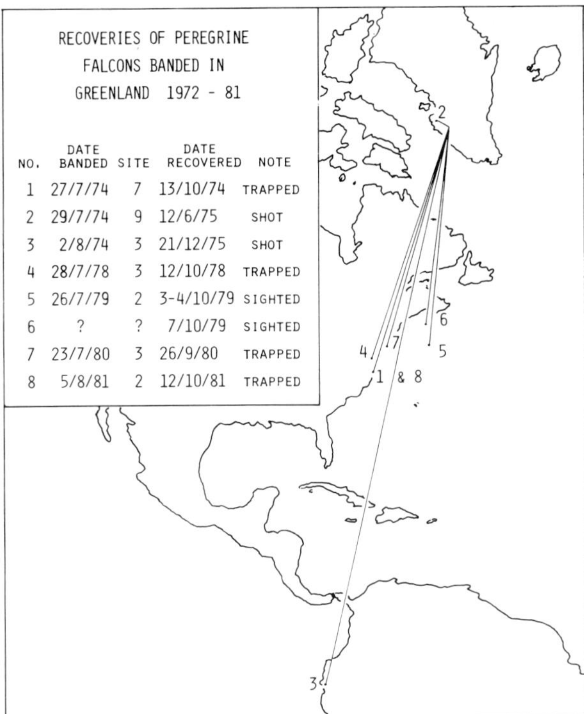

NO.	DATE BANDED	SITE	DATE RECOVERED	NOTE
1	27/7/74	7	13/10/74	TRAPPED
2	29/7/74	9	12/6/75	SHOT
3	2/8/74	3	21/12/75	SHOT
4	28/7/78	3	12/10/78	TRAPPED
5	26/7/79	2	3-4/10/79	SIGHTED
6	?	?	7/10/79	SIGHTED
7	23/7/80	3	26/9/80	TRAPPED
8	5/8/81	2	12/10/81	TRAPPED

Fig. 12. This map shows recoveries for peregrine falcons banded in Greenland.

least two main routes from North America to South America. One is the usual Atlantic coastal flyway to Florida, USA, thence southeast along the Caribbean Islands. In the second route, small birds leave the Atlantic east coast (anywhere from Nova Scotia, Canada, to Virginia, USA) and fly southeast over the ocean. This movement has been confirmed by numerous observations (radar and shipboard), at least as far as the area of the Sargasso Sea. From there, the masses of small birds apparently take advantage of the northeast trade winds to carry them in a southwesterly direction over the Caribbean to South America. The autumn peregrine migration in eastern North America certainly has been observed along the Atlantic coast. After Holcomb's single sighting aboard ship 200 miles at sea, and several other reports from sea captains, part of the autumn migration of this raptor seems to follow the same path as the oceanic movement of small birds.

Several days after Holcomb's sighting on Belagorsk, P. Willis observed a red-banded peregrine at Cape Sable, Nova Scotia, Canada (no. 6). Poor light conditions prevented him from determining plumage color (age) of this bird, and he could not read band numbers.

Two additional recoveries (nos 7 & 8) fall along the east coast migration path: a male banded on 23 July 1980 at the same cliff as recoveries nos 3 and 4 above, was trapped and released by banders at Democrat Point, New York, USA, on 26 September 1980 (C. Safina pers. comm.). A female banded on 5 August 1981 at the same cliff as the Belagorsk falcon was trapped and released by bird banders at Fisherman's Island, Virginia, USA, on 12 October 1981 (M. Bird pers. comm.). This location is within sight of the first recovery location (Cape Charles, Virginia, USA) in 1974.

Gyrfalcon

No gyrfalcons had been banded in Greenland before 1946. Between 1946 and 1965, 51 gyrfalcons were banded there, five of which were later recovered, all in Greenland (Mattox 1970a). These returns confirm references in the literature which state that the gyrfalcon in Greenland is either non-migratory or it moves in a general southerly direction within Greenland in winter.

In 1967, W. Mattox and R. Graham banded fourteen gyrfalcons in autumn at Disko, ten females and four males (Mattox 1970b). Of the fourteen falcons banded, two (14%) were later recovered, both females. One gyrfalcon was found dead 390 km south-southwest of the banding location four weeks after banding. The other gyrfalcon was shot seven weeks after banding 160 km south-southwest of the banding location. These two recoveries agree with the above statement about a general southerly movement in West Greenland.

From 1972 to 1981 we banded 23 gyrfalcons, 15 in the survey area and eight at the autumn trapping location near Frederikshåb. No recoveries have been received from these recent bandings.

Acknowledgements

This project, spanning ten field seasons, involving 41 field researchers, and relying on numerous sources of support in the United States and Greenland, could not have been accomplished by the two authors alone. First, we owe sincere thanks to the hard work under difficult conditions by the participants. Secondly, the authors and participants are eternally grateful to the many American, Danish, and Greenlandic friends who helped with the project and made it possible.

Travel to Greenland was supported by the U.S. Air Force (1972), and from 1973 by the U.S. Army Chemical Systems Laboratory, Ecology Branch (F. P. Ward, Chief). We acknowledge major support from Dartmouth College (Mellon Fund), American Museum of Natural History (Frank M. Chapman Fund), University of Southern Colorado (G. Farris and J. Linam), Brundage Foundation, Institute of Current World Affairs-Crane Foundation, The Explorers Club, Holubar Mountaineering, Inc., Brigham Young University (C. White), The Peregrine Fund of Cornell University, and The Peregrine Fund, Inc. (T. Cade, President).

We emphasize particularly the support and interest of the late Finn Salomonsen and F. Prescott Ward, without whose patience and encouragement this lengthy project would have been impossible.

James Enderson, Tom Cade, and Clayton White reviewed the drafts of the manuscript, a very valuable and time-consuming contribution. Pat Burnham kindly typed various drafts of this paper. Karen Greiner and Wanda Wedemeyer also provided valuable typing assistance. Finally, and very importantly, we wish to sincerely thank our families for their tolerance during all of our long absences and their personal support and interest in our work.

References

Anderson, D. W., & Hickey, J. J. 1972. Eggshell changes in certain North American birds. – Proceedings of the XVth International Ornithological Congress, The Hague, 1970. – E. J. Brill, Leiden: 514–540.

Berger, D. D., Anderson, D. W., Weaver, J. D., & Risebrough, R. W. 1970. Shell thinning in eggs of Ungava peregrines. – Can. Field-Nat., 84 (3): 265–267.

Berger, D. D. & Mueller, H. C. 1969. Nesting peregrine falcons in Wisconsin and adjacent areas. – In: Hickey, J. J. (ed.), Peregrine Falcon Populations, their biology and decline. – Univ. Wisconsin Press, Madison: 115–122.

Bertelsen, A. 1932. Meddelelser om nogle af de i Vestgrønlands distrikter mellem 60° og 77° n.br. almindeligere forekommende fugle, særlig om deres udbredelsesområde, deres yngleområde og deres træk. – Meddr Grønland, 91 (4): 29 pp.

Bond, R. M. 1946. The peregrine population of western North America. – Condor, 48 (3): 101–116.

Burnham, W. A. 1975. Breeding biology and ecology of the peregrine falcon *(Falco peregrinus)* in West Greenland. – M.S. thesis, Brigham Young Univ., unpubl. 57 pp.

Burnham, W. A., Craig, J., Enderson, J. H., & Heinrich, W. R. 1978. Artificial increases in reproduction of wild peregrine falcons. – J. Wildl. Manage., 42 (3): 625–628.

Burnham, W. A., Mattox, W. G., Jenkins, M. A., Clement, D. M., Harris, J. T., & Ward, F. P. 1974. Falcon research in Greenland, 1973. – Arctic, 27 (1): 71–74.

Cade, T. J. 1960. Ecology of the peregrine and gyrfalcon populations in Alaska. – Univ. Calif. Publs Zool., 63 (3): 151–290.

Cade, T. J., Lincer, J. L., White, C. M., Roseneau, D. G., & Swartz, L. G. 1971. DDE residues and eggshell changes in Alaska falcons and hawks. – Science, 172: 955–957.

Dementiev, G. P. 1960. Der Gerfalke. – (Neue Brehm Bücherei, no. 264). – Wittenberg: A. Ziemsen Verlag. 88 pp. (The Gyrfalcon, translation by Dept. of the Sec. of State, Foreign Languages Div., Govt. of Canada. 142 pp. multilith.).

Enderson, J. H., & Craig, J. 1974. Status of the peregrine falcon in the Rocky Mountains in 1973. – Auk, 91 (4): 727–736.

Fencker, E. C., & Scheel, H. 1929. Nogle efterladte Optegnelser om Nordgrønlands Fugle. – Dansk Ornithologisk Foren. Tidsskr. 23: 47–48.

Fischer, W. 1968. Der Wanderfalk *(Falco peregrinus* and *Falco pelegrinoides).* Die Neue Brehm Bücherei, Berliner Tierpark-Buch Nr. 13. (The peregrine falcon, translation by Dept. of the Sec. of State, Foreign Languages Div., Govt. of Canada. typescript).

Fyfe, R. 1969. The peregrine falcon in the Canadian arctic and eastern North America. – In: Hickey, J. J. (ed.), Peregrine Falcon Populations, their biology and decline. – Univ. Wisconsin Press, Madison: 101–114.

Fyfe, R. W., Temple, S. A., & Cade, T. J. 1976. The 1975 North American peregrine falcon survey. – Can. Field-Nat., 90 (3): 228–273.

Hagar, J. A. 1969. History of the Massachusetts Peregrine Falcon, 1935–57. – In: Hickey, J. J. (ed.), Peregrine Falcon Populations, their biology and decline. – Univ. Wisconsin Press, Madison: 123–131.

Harris, J. T., & Clement, D. M. 1975. Greenland peregrines at their eyries. – Meddr Grønland, 205 (3): 1–28.

Hays, H., & Risebrough, R. 1972. Pollutant concentrations in abnormal young terns from Long Island Sound. – Auk, 89 (1): 19–35.

Herbert, R. A., & Herbert, K. G. 1969. The extirpation of the Hudson River peregrine falcon population. – In: Hickey, J. J. (ed.), Peregrine Falcon Populations, their biology and decline. – Univ. Wisconsin Press, Madison: 133–154.

Hickey, J. J. 1942. Eastern populations of the duck hawk. – Auk, 59 (2): 176–204.

– (ed.), 1969. Peregrine Falcon Populations, their biology and decline. – Univ. Wisconsin Press, Madison: xxii + 596 pp.

Hickey, J. J., & Anderson, D. W. 1969. The peregrine falcon: life history and population literature. – In: Hickey, J. J. (ed.), Peregrine Falcon Populations, their biology and decline. – Univ. Wisconsin Press, Madison: 3–42.

Hickey, J. J., & Roelle, J. E. 1969. Conference summary and conclusions. – In: Hickey, J. J. (ed.), Peregrine Falcon Populations, their biology and decline. – Univ. Wisconsin Press, Madison: 553–567.

Jenkins, M. A. 1978. Gyrfalcon nesting behavior from hatching to fledging. – Auk 95 (1): 122–127.

Mattox, W. G. 1970a. Bird-banding in Greenland. – Arctic 23 (4): 217–228.

– 1970b. Banding gyrfalcons *(Falco rusticolus)* in Greenland, 1967. – Bird-Banding 41 (1): 31–37.

– 1975. Bird of prey research in West Greenland, 1974. – Polar Rec. 17 (109): 387–388.

Mattox, W. G., Graham, R. A., Burnham, W. A., Clement, D. M., & Harris, J. T. 1972. Peregrine falcon survey, West Greenland, 1972. – Arctic 25 (4): 308–311.

Newton, I., & Bogan, J. 1974. Organochlorine residues, eggshell thinning, and hatching success in British sparrow hawks. – Nature, London 249 (5457): 582–583.

Oldendow, K. 1933. Fugleliv i Grønland. – Grønl. Selsk. Årsskr. 1932–33: 147–151.

Peakall, D. B., & Peakall, M. L. 1973. Effect of a polychlorinated biphenyl on the reproduction of artificially and naturally incubated dove eggs. – J. Appl. Ecol. 10: 863–868.

Porter, R. D., & White, C. M. 1973. The peregrine falcon in Utah, emphasizing ecology and competition with the prairie falcon. – Bull. Brigham Young Univ., Biol. Serv., XVIII (1): 1–74.

Prestt, I. 1965. An enquiry into the recent breeding status of some of the smaller birds of prey and crows in Britain. – Bird Study 12: 196–221.

Ratcliffe, D. A. 1969. Population trends of the peregrine falcon in Great Britain. – In: Hickey, J. J. (ed.), Peregrine Falcon Populations, their biology and decline. – Univ. Wisconsin Press, Madison: 239–269.

– 1970. Changes attributable to pesticides in egg breakage frequency and eggshell thickness in some British birds. – J. Appl. Ecol. 7 (1): 67–115.

– 1972. The peregrine populations of Great Britain in 1971. – Bird Study 19: 117–156.

– 1980. The Peregrine falcon. – Buteo Books, Vermillion, S. D.: 416 pp.

Rice, J. N. 1969. The decline of the peregrine population in Pennsylvania. – In: Hickey, J. J. (ed.),. Peregrine Falcon Populations, their biology and decline. – Univ. Wisconsin Press, Madison: 155–163.

Ryves, B. H. 1948. Bird life in Cornwall. – Collins, London: 256 pp.

Salomonsen, F. 1950–51. Grønlands Fugle/Birds of Greenland. – København, Ejnar Munksgaard: 607 pp.

– 1967. Fuglene på Grønland. – København, Rhodos: 340 pp.

Schiøler, E. L. 1925–31. American peregrine falcon, *Falco peregrinus anatum*. – In: Danmarks Fugle: 399–405. (transl. from Danish)

Walker, W., Mattox, W. G., & Risebrough, R. W. 1973. Pollutant and shell thickness determinations of peregrine eggs from West Greenland. – Arctic 26 (3): 255–256.

White, C. M. 1968. Diagnosis and relationship of the North American tundra-inhabiting peregrine falcon. – Auk 85: 179–191.

White, C. M., & Cade, T. J. 1971. Cliff-nesting raptors and ravens along the Colville River in arctic Alaska. – The Living Bird, 10th Annual: 107–150.

White, C., & Risebrough, R. 1977. Polychlorinated biphenyls in the ecosystems. – In: Murit, M. L., & Fuller, R. G. (eds.), The environment of Amchitka Island, Alaska. – U.S. ERDA, TID-26712, Oak Ridge, Tenn.: 615–625.

Williams, T. C., & Williams, J. M. 1978. An oceanic mass migration of land birds. – Scient. Am. 239 (4): 166–176.

Winge, H. 1898. Grønlands Fugle. – Meddr Grønland 21: 247–249.

1980

3. H. Meltofte, M. Elander and C. Hjort:
»Ornithological observations in Northeast Greenland between 74°30′ and 76°00′ N. lat. 1976«. 53 pp.

The results of one summer's work in central Northeast Greenland are presented. The avifauna in the country traversed on several extensive survey trips is described. More intensive studies were made in an 18.2 km² census area on southernmost Hochstetter Forland. Here the populations were followed throughout the breeding season, and information on arrival, pre-laying period, population densities, habitat and nest site selection, breeding schedule, clutch size, hatching success, re-nesting, non-breeders, moult, post-breeding activities and departure is given. Special attention is given to *Clangula hyemalis, Somateria spectabilis, Anser brachyrhynchus, Arenaria interpres, Calidris maritima, Calidris alpina, Calidris alba, Phalaropus fulicarius* and *Stercorarius longicaudus*. An extremely high predation pressure was caused by *Alopex lagopus,* and this is discussed in relation to lemming abundance and environmental conditions.

1981

7. J. de Korte, C. A. W. Bosman & H. Meltofte:
»Observations on waders (Charadriidae) at Scoresby Sund, East Greenland«. 21 pp.

Populations of waders in three census areas at Scoresby Sund, central East Greenland, were studied during the three breeding seasons of 1973, 1974 and 1975. Ringed Plover (*Charadrius hiaticula*), Golden Plover (*Pluvialis apricaria*), Turnstone (*Arenaria interpres*), Knot (*Calidris canutus*), Dunlin (*Calidris alpina*) and Sanderling (*Calidris alba*) bred in the census areas, while Purple Sandpiper (*Calidris maritima*) and Red-necked Phalarope (*Phalaropus lobatus*) bred elsewhere in the region. Population densities were very low, compared to other areas further north in high arctic Greenland. Extensive, deep and late-thawing snow cover prevents waders from utilizing large areas in June. Time of breeding showed a high correlation with the snow melting conditions in the respective areas and years. Breeding success was generally low; only Ringed Plover had more than 50% nest and egg survival. Nest failures were probably mostly due to predation by Arctic Foxes (*Alopex lagopus*). Observation and examination of individuals from post-breeding flocks in the second half of July indicated that these flocks contained mainly non-breeders, but failed and successful breeders were also present. Measurements on eggs, pulli and adults are presented.

1982

8. Helge Abildhauge Thomsen:
»Planktonic choanoflagellates from Disko Bugt, West Greenland, with a survey of the marine nanoplankton of the area«. 35 pp.

Light and electron microscopy of whole mounts prepared from water samples collected in July and August 1977 at thirteen stations in the vicinity of Godhavn (Disko Bugt, West Greenland), has led to the enumeration of approximately 100 nanoplanktonic taxa. A full account is given of field and laboratory methods. The most conspicuous algal class was the Prymnesiophyceae with more than 38 species. Among the heterotrophic organisms listed the Choanoflagellida was the most important single group, comprising 28 species. Two new choanoflagellate taxa are described on the basis of West Greenland material: *Conion groenlandicum* gen. et sp.nov. and *Diaphanoeca undulata* sp.nov.

In order to facilitate immediate comparison of closely related taxa *Diaphanoeca sphaerica* sp.nov. is described on the basis of Danish material.

Thirteen of the loricate choanoflagellate species listed are new recordings for West Greenland. A summary of previous findings of the choanoflagellate species encountered in the Disko Bugt samples show that three species *(Conion groenlandicum, Pleurasiga caudata* and *Parvicorbicula serratula)* are so far known from arctic and subarctic localities only. A pronounced vertical distribution pattern of choanoflagellate species was observed at one station southeast of Godhavn. Three distinct species associations occurred in this particular water column (0–300 m).

1982

9. Eric Steen Hansen:
»Lichens from Central East Greenland«, 33 pp.

A total of 600 samples of 167 species of macro- and microlichens were collected mainly by Pauline Topham and Geoffrey Halliday on botanical expeditions to Central East Greenland in the years 1961, 1962, 1968, 1971, 1974 and 1980. Three of the species, viz., *Caloplaca tornoensis* Magnusson, *Rhizocarpon pusillum* Runem, and *Verrucaria thalassina* (Zahlbr.) Zsch. are additions to the known lichen flora of Greenland. The following eleven species have not previously been reported from East Greenland: *Catillaria philippea* (Mont.) Massal., *Cladonia luteoalba* A. Wilson & Wheldon, *C. macroceras* (Delise) Ahti, *Coelocaulon divergens* (Ach.) R. H. Howe, *Diploschistes muscorum* (Scop.) R. Sant., *Leprocaulon subalbicans* (Lamb) Lamb & Ward, *Peltigera kristinssonii* Vitik., *Pertusaria octomela* (Norman) Erichsen, *Rhizocarpon intermediellum* Räsänen, *Solorina saccata* (L.) Ach. and *Thelidium papulare* (Fr.) Arnold.

Information is provided on climatic conditions at two meteorological stations situated in the area investigated. Thirtyeight collecting localities are listed, together with brief notes on their geology. The localities are situated between the southernmost part of Liverpool Land and Jameson Land, c. 70°N, and the middle of Lyell Land and Traill Ø, c. 73°N.

A survey is given of some important ecological, phytosociological and distributional characteristics for the lichen species, together with information on the presence of perithecia or apothecia.

Lichens of particular interest are discussed in the special part of the paper.

A number of commonly used synonyms are listed in the Appendix.

1982

10. F. J. A. Daniëls:
»Vegetation of the Angmagssalik District, Southeast Greenland, IV. Shrub, dwarf shrub terricolous lichens«, 78 pp.

This paper deals with part of the results of the Dutch phytosociological expeditions in 1968 and 1969 to the Angmagssalik District, Southeast Greenland.

Shrub, dwarf shrub and terricolous lichen vegetation is treated here. The general part contains a description of the Angmagssalik District with emphasis on the applied methods.

The vegetation has been studied according to concepts of the French–Swiss School. The typology is based on about 250 records. The procedure of differentiation and classification of the plant communities is discussed. The term "decisive" differential taxon is introduced and defined. The association concept is considered from a regional point of view. The plant communities are arranged in a floristic hierarchic system.

Concerning habitat factors, the altitude a.s.l., slope and wind direction were measured. Other factors were roughly estimated. The soil types are indicated.

The following part contains a discussion of the vegetation units, with their floristic composition and physiognomy, habitat and distribution, and syntaxonomic position. This includes 24 vegetation units, 1 complex of communities, 11 communities and 12 associations. These are designed to the classes Oxycocco–Sphagnetea, Scheuchzerio–Caricetea, Betulo–Adenostyletea, Loiseleurio–Vaccinietea, Carici–Kobresietea, Salicetea herbaceae and Juncetea trifidi. Eight new associations and 1 new alliance are presented. Some syntaxa have been revised or validated.

The classification by Molenaar (1976) of mire vegetation and chionophytic herb communities is discussed and a new classification is proposed. Dwarf shrub vegetation with *Empetrum hermaphroditum* and/or *Vaccinium microphyllum* on acid, mainly mineral soil is extremely varied in composition and physiognomy and is considered a zonal formation, which largely determines the aspect of the region. The *Empetrum–Vaccinium* community is the climax vegetation of the district.

The greater part of the communities and associations can be assigned to alliances described from Scandinavia, and the phytosociological relationship with that region is emphasized. Only the Dryadion integrifoliae and the Cladonio–Viscarion all. nov. are not known from Scandinavia. The vegetation of the Angmagssalik District has its own character, as shown on the association level by the Sphagno–Salicetum, the Rhododendro–Vaccinietum, the Gymnomitrio–Loiseleurietum, the Carici–Dryadetum and the Cladonio–Viscarietum (all new), which are actually restricted to the area. The other 7 associations are also found at the southern and western coasts of Greenland. Most vegetation types (associations and communities) have a lowarctic-oceanic distribution. A few types are also found in Iceland and Scandinavia.

1983

11. Tyge W. Böcher:
»The allotetraploid *Saxifraga nathorsti* and its probable progenitors *S. aizoides* and *S. oppositifolia.*« 22 pp.

Saxifraga nathorsti is an endemic Greenland species geographically restricted to Northeast Greenland. Morphologically it is intermediate between *Saxifraga oppositifolia* with purplish petals and *S. aizoides* with yellow petals. A hybrid between these two species is difficult to obtain and is not known from Greenland or anywhere else.

New material from Northeast Greenland has been cultivated and studied cytologically. One strain of typical *S. nathorsti* corresponded to the material studied previously. It also had 52 chromosomes and showed a high degree of pairing during meiosis. It was fertile, but exhibited several meiotic irregularities. Another strain seemed morphologically more closely related to *S. oppositifolia.* It was sterile and had the triploid number 2n = 39. It was assumed to have two genomes from *S. oppositifolia* and one from *S. aizoides.* It appears most probable that triploids of this kind after fertilization with pollen from *S. aizoides* can give rise to *S. nathorsti.*

Anatomical studies of the structure of epithem hydathodes in *S. nathorsti* and its two possible ancestors, *S. oppositifolia* and *S. aizoides,* show that *S. nathorsti* in several important hydathode characters occupies an intermediate position between *S. oppositifolia* and *S. aizoides.* Thus, all available facts support the theory of the origin and stabilization of *S. nathorsti* as an allotetraploid species.

1984

12. Ole G. Norden Andersen:
»Meroplankton in Jørgen Brønlund Fjord, North Greenland.« 25 pp.

Meroplanktonic larvae of at least 41 species of bottom invertebrates in Jørgen Brønlund Fjord, North Greenland (82°10′N, 30°30′W) are described with respect to species identification, occurrence, reproduction, development, growth, settlement, and relations to depth, light, hydrography, and primary production. A few holoplankters and some "pseudoplanktonic" nematodes are included. The occurrence of such a large number of species with pelagic larvae does not invalidate "Thorson's rule" of 1950, stating that the number of species having pelagic larval development decreases, as one moves from the equator to the pole, but it does lead to a less strict interpretation of it. Several species have lecithotrophic pelagic development. The short period of primary production, however meager, seems vital to many of the planktotrophic larvae, in promoting growth and settling, although the spawning of many species is not stricktly linked to this period. Larvae of *Hiatella striata* (Fleuriau) seem able to live in the plankton for a year, even surviving the long dark, winter.

13. Thomas K. Kristensen:
»Biology of the squid Gonatus fabricii (Lichtenstein, 1818) from West Greenland waters.« 17 pp.

Three hundred adult and subadult *Gonatus fabricii* and about 7000 juveniles from West Greenland waters were examined. In spring and early summer large numbers of juvenile *G. fabricii* hatch in Davis Strait. Their abundance fluctuates from year to year. In Disko Bugt the juveniles hatch in autumn and early winter. Juvenile *G. fabricii* hatch over a large area in Davis Strait at depths exceeding 200 m. At night juveniles south of the polar circle perform vertical upward migrations. Likewise it seems that shoals of juveniles disperse at the same time. The number of juvenile *G. fabricii* is found to be about the same as the number of larvae of the Greenland halibut, *Rheinhardtius hippoglossoides,* a common commercial fish. The growth of *G. fabricii* was found to be 8–9 mm per month. The development of the gonads in relation to pen length is describable by the allometric equation. The testis begins to develop at a pen length of about 8–10 cm, the penis at a pen length of 3–5 cm. The largest mature male measured 29.3 cm pen length. The ovary begins to develop at a pen length of 6–8 cm. No mature females were found. In Greenland waters males probably mature at about 20 cm pen length, females between 25 and 30 cm pen length.

51% of specimens had empty stomachs, 27% were half full and 22% full. Crustaceans, fish and cephalopods were found in the stomachs and crustaceans were the most important. The protein percent was found to be 12.5 and in the liver the lipid percent was 63.

Spawning and predators of *G. fabricii* are also discussed.